Lance Douglas

Five Bizzy Honeybees

Snow Canyon Publishing

The long, cold winter is over and the bright spring sunshine begins to warm the home of the five little honey bees. The sweet smelling flowers outside cause the five little honey bees to dance with excitement.

The traditional dome-shaped beehive, called a skep, was replaced by the wooden box beehive, with removable honey frames, in the mid-1800's. However, the timeless skep remains the symbol of the beehive.

Inside the hive their mother, the queen bee, declares, "It's springtime! The flowers are blooming and you know what that means. Now get busy my five little honey bees." The five little honey bees quickly get to work. Each bee knows what she must do to help make delicious honey.

A colony of bees is highly organized, containing one queen and up to 100,000 worker bees. Worker bees are undeveloped females who instinctively carry out their specific duties throughout their lifespan.

The first little honey bee quickly flies out the door and searches for the blossoms with the sweetest nectar. She will carry the nectar back to the hive where it will be turned into honey.

Honey bees are attracted to blossoms containing the sweetest nectar rather than the brightest colors. While on nectar-rich blossoms, she sucks the nectar through her tube-like tongue and fills her special "honey stomach." She will return to the hive where the nectar will be transferred to other worker bees who will add enzymes to the nectar until it is suitable for storage as honey in the honeycomb.

The second little honey bee also quickly flies out of the hive, but she is in search of the blossoms with the richest yellow pollen. She will load her pollen baskets to be so heavy that she can barely fly back to the hive.

Honey bees have special "pollen baskets" on their back pair of legs. Pollen is loaded into these baskets and carried back to the hive to be deposited in the cells.

Back at the hive the second honey bee will place the pollen into the wax cells to be used as food for the growing baby bees.

After filling their baskets with pollen, the pollen is delivered to the hive and stored in cells near the developing baby bees (larvae). The pollen is the primary source of protein for the developing bees.

The third little honey bee follows her sisters out the door. However, she is not searching for nectar or pollen; she is looking for cool, clear water. She will carry the water back to the hive to be shared with her sisters and to help keep the hive cool during the hot summer days.

Like all living things, honey bees need a drink of water on a hot summer day. Additionally, water is sprinkled in tiny droplets throughout the hive, creating an evaporative cooling system as the bees circulate air through the hive, maintaining a steady temperature within the hive.

The fourth little honey bee stays inside the hive to help her mother, the queen. The queen will be laying many eggs every day so there can be new bees to help with the work of the hive. The fourth little bee will feed the pollen to the growing baby bees. She also helps to keep the hive clean and neat.

Designated "nurse bees" are responsible for feeding the developing bee larvae. The protein found in pollen is essential to this process.

The fifth little honey bee will also stay at home. She is stronger than her sisters, so she will stand by the door to make sure no unwanted creatures get into the hive and steal their precious honey. While she stands guard, she will fan her wings to blow air through the hive to keep it cool.

Beginning at the hive entrance, and throughout the hive, strategically positioned bees fan their wings to create airflow through the hive. Additionally, the hive entrance has a small army of "guard bees" who patrol the entrance. They sound an alarm at the first sign of trouble, alerting the rest of the colony to go on the attack.

From sunup to sundown the five little bees work tirelessly to fill their hive with honey. When the summer is over and the leaves start to fall, the five little bees share their honey with their friend the beekeeper. The beekeeper will be sure to leave enough honey in the hive to last the hungry bees through the winter.

When harvesting the honey from the beehive, a beekeeper must leave enough honey to last the bees through the cold winter months. This will be their only source of food until spring.

The beekeeper carefully takes the honey out of the honeycomb. He fills many jars with honey and returns the empty honeycomb to the beehive. The bees are excited for spring so they can start working again..........

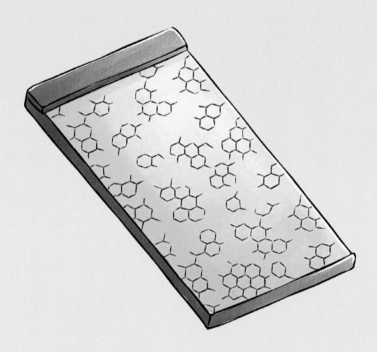

The process of removing the honey from the comb is called "extracting." The honeycomb is "uncapped" with a hot knife and then rapidly spun in an "extractor," causing the honey to flow out of the cells.

But for now it is time to rest.

During the cold winter months bees will "cluster" tightly together and create a vibrating motion to generate the heat necessary to maintain the temperature inside the hive.

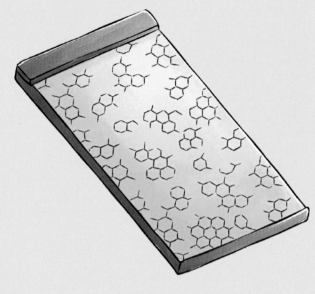

A single bee will produce only 1/12 teaspoon of honey in her lifetime.

During the busy "nectar flow" months, worker bees live approximately six weeks. During her six week lifespan, a worker bee will progress from one duty to another, beginning at nurse bee and completing her life by foraging for water, nectar and pollen.

Honey bees will travel up to three miles from their hive to collect water and to forage for nectar and pollen.

un – BEE - lievable!

Honey bees are the only insects that produce food for humans.

A typical beehive can produce up to 400 pounds of honey per year.

Honey is the only food that includes all the substances necessary to sustain life, including water.

To make one pound of honey, the worker bees in the colony must visit two million flowers, fly over 55,000 miles and will take the lifetime work of approximately 300 bees.

Bees communicate with each other inside the hive by dancing in a figure eight pattern. The dance communicates to the other bees where to find water and rich sources of nectar and pollen. The dance provides direction and distance in relation to the position of the sun in the sky at that moment.

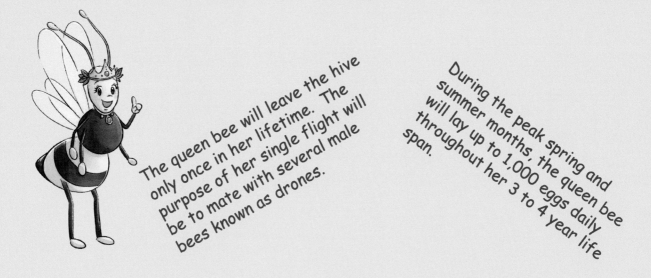

The queen bee will leave the hive only once in her lifetime. The purpose of her single flight will be to mate with several male bees known as drones.

During the peak spring and summer months, the queen bee will lay up to 1,000 eggs daily throughout her 3 to 4 year life span.

Hail to the Queen

In a box hive, the bottom box (brood nest) is reserved for the queen to lay her eggs, and for pollen storage. Honey is stored in the upper boxes.

Male honey bees, called drones, are much larger than worker bees. Their sole purpose is to mate with a queen once in her lifetime. Drones are permitted to remain in the hive in case the queen dies and a new queen is created. Otherwise, drones have no purpose. Most will be driven from the hive in the winter.

Worker bees can turn a normal egg that has been laid by the queen into a queen bee by feeding it "royal jelly." As the queen bee ages and her egg production decreases, the worker bees sense this decrease and they create a new queen who will leave the hive only once to mate, and the cycle begins again.

The miraculous life of the honey bee
has brought wonder, awe, amazement and joy
into my life, and the life of my family. My hope
is that it can do the same for you.

Lance Douglas - *Author*

Made in the USA
San Bernardino, CA
24 April 2018